JUPITER

JUPITER

SEYMOUR SIMON

WILLIAM MORROW AND COMPANY, INC.
New York

PHOTO CREDITS

All photographs courtesy of NASA, except page 7,
courtesy of Kyle Cudworth, The Yerkes Observatory.

Inquiries should be addressed to William Morrow and
Company, Inc., 105 Madison Avenue, New York, N.Y. 10016.
Printed in Verona, Italy.

1   2   3   4   5   6   7   8   9   10

Library of Congress Cataloging in Publication Data
Simon, Seymour.   Jupiter.
Summary: Describes the characteristics of the planet Jupiter
and its moons as revealed by photographs sent back by
two unmanned Voyager spaceships which took
one-and-one-half years to reach this distant giant.
1. Jupiter (Planet) — Juvenile literature.
[1. Jupiter (Planet) 2. Planets]   I. Title.
QB661.S585 1985   523.4'5   85-2922
ISBN 0-688-05796-9
ISBN 0-688-05797-7 (lib. bdg.)

To Robert and Nicole

←Jupiter

From our planet Earth, Jupiter looks like a bright
star in the night sky. But Jupiter is a planet, too.

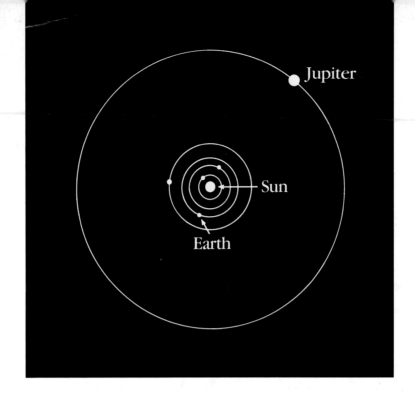

Nine planets travel around the sun. The sun and its planets are called the Solar System. Earth is the third closest planet to the sun. Jupiter is the fifth planet — it is about 480 million miles from the sun. Jupiter is so far away in space it takes almost 12 years to go around the sun once.

In August and September of 1977, two unmanned *Voyager* spaceships were launched from Earth to get a close-up view of Jupiter. Traveling at great speeds, the spaceships still took one-and-one-half years to reach the distant planet. This photograph was taken when *Voyager 1* was still 32 million miles from Jupiter.

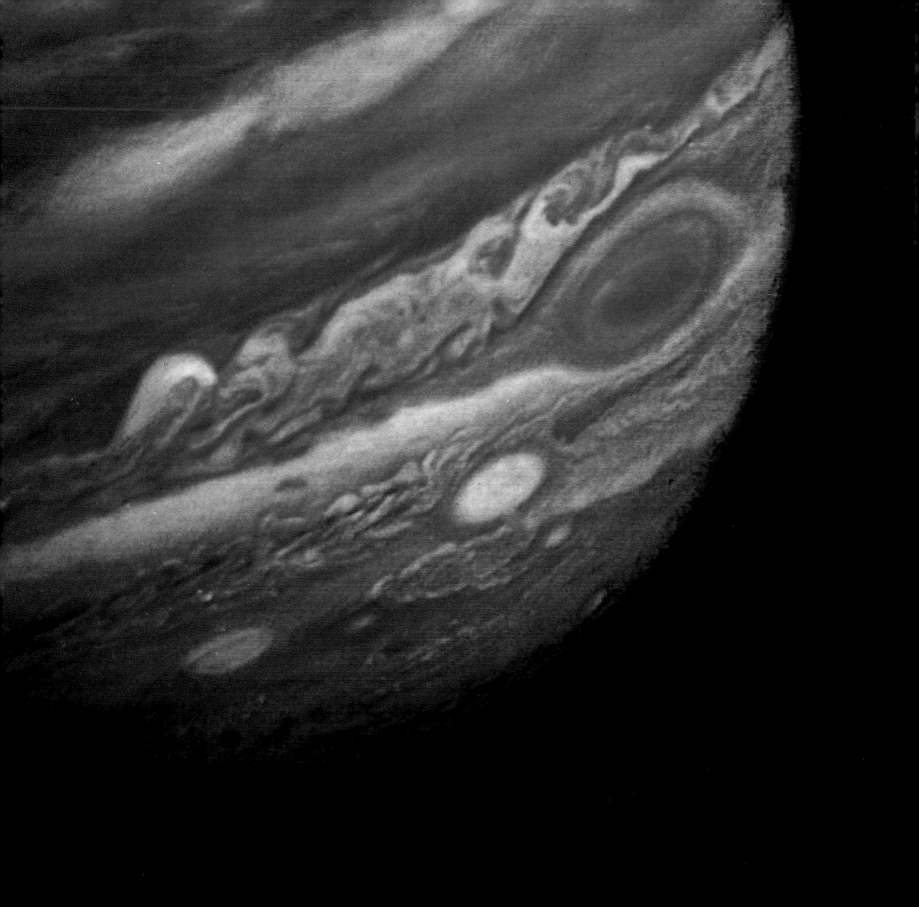

Jupiter is the giant planet of the Solar System. It is more than one-and-one-half times as big as the other eight planets put together. If Jupiter were hollow, more than 1300 planet Earths could fit inside.

No one has ever seen the surface of Jupiter. The planet is covered by clouds hundreds of miles thick. We can see only the tops of the clouds: bands of reds, oranges, tans, yellows, and whites.

Jupiter's clouds are not made of tiny water droplets like clouds on Earth. Jupiter's atmosphere is mostly hydrogen gas. The hydrogen and small amounts of other gases form the brightly colored clouds.

The clouds are always moving. Look at the large white cloud at the bottom of the photographs. In four months, it has moved thousands of miles to the east *(right)*.

Our planet Earth turns around once every twenty-four hours. Even though Jupiter is much larger, it spins around in less than ten hours. That's the fastest of any planet. This rapid spin causes powerful winds that push the clouds into colorful bands, streaks, and swirls that circle the planet.

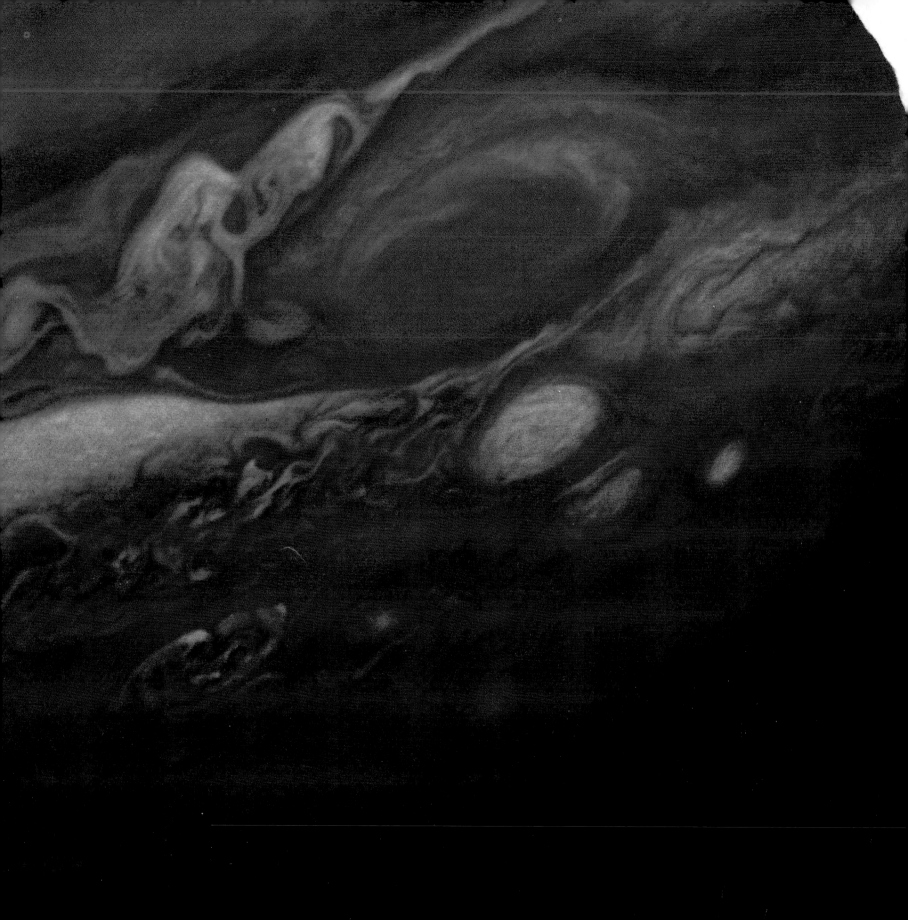

One of the many mysteries on Jupiter is the Great Red Spot. The spot was first seen through a telescope from Earth more than three hundred years ago. But no one knows when it first formed. The giant spot is probably an enormous storm, a super-hurricane more than twice as big as our whole Earth.

The Great Red Spot has changed through the years. Sometimes it shrinks and becomes a dull pink. Other times it grows and becomes bright red. But unlike other clouds on Jupiter, the Great Red Spot does not change its position and it has kept the same oval shape for centuries.

Imagine exploring Jupiter from a spaceship. As you look down at the tops of the clouds, the air has the same light blue color that Earth's sky has in the daytime. The temperature here is very cold—over 250 degrees (F) below freezing.

The colors of the clouds change as you fly lower. The upper clouds are mainly white and blue. The lower clouds are orange, yellow, and brown, and the temperature is warmer here. It is dark outside your spaceship, as little or no sunlight filters down through the clouds. Below you, gigantic bolts of lightning flash across the sky and light up the darkness.

Jupiter's "surface" is an ocean of liquid hydrogen that covers the entire planet. This ocean may be ten thousand or more miles deep—no one knows. Perhaps Jupiter has no solid surface at all but is entirely liquid down to its rocky center, almost ninety thousand miles below the clouds.

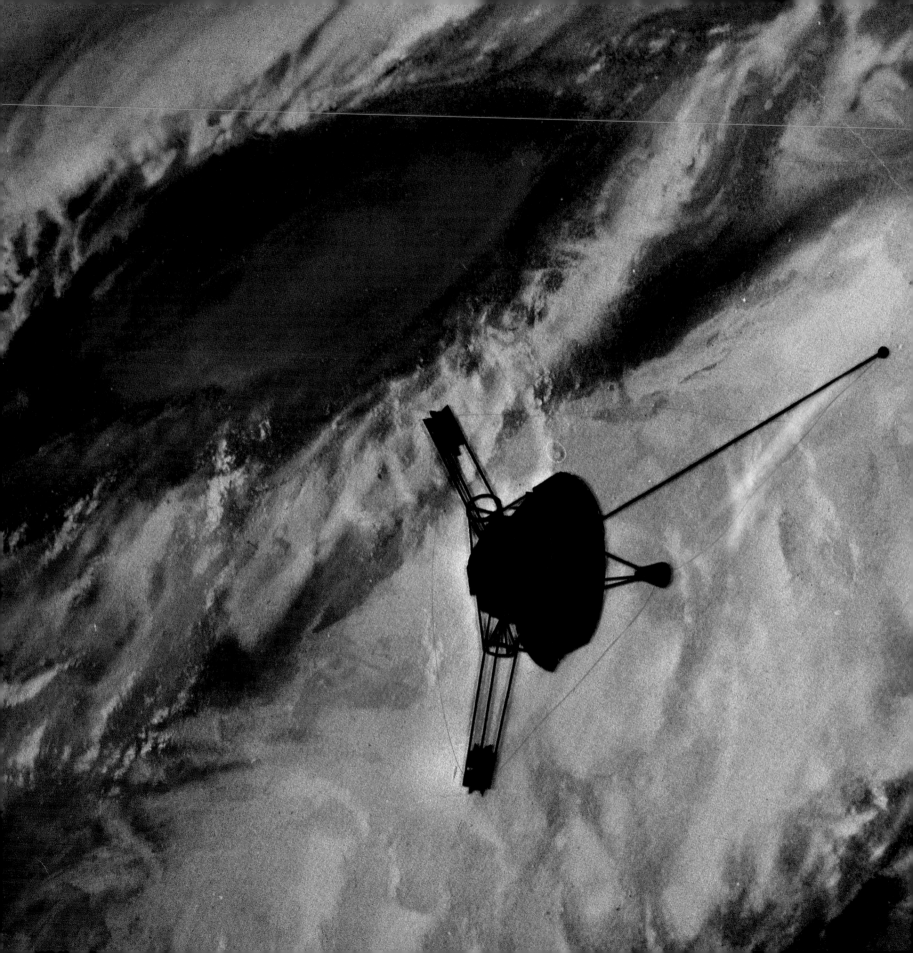

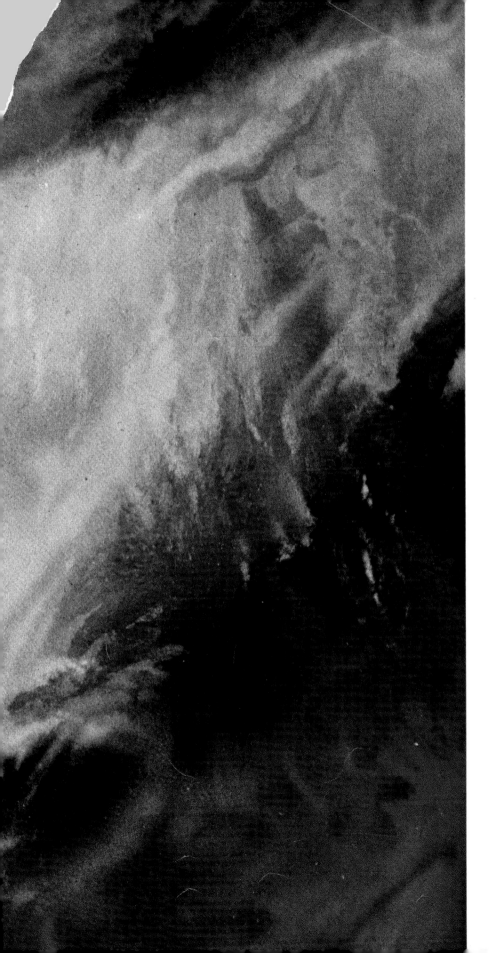

Jupiter is very hot at its center—more than 50,000 degrees (F). Like a boiling pot of water, the heat from below stirs up the clouds so that they rise and sink.

If you were on the surface of Jupiter, you would weigh more than two-and-one-half times more than what you weigh on Earth. If you weigh 100 pounds on Earth, you would weigh 264 pounds on Jupiter.

Jupiter has at least sixteen moons and more may still be discovered. The outer moons are small, most under 50 miles across. The four largest moons circle close to the planet: Io, Europa, Ganymede, and Callisto. These are called the Galilean moons, after their discoverer, the great Italian scientist Galileo. Galileo first saw the moons in 1610 with his small homemade telescope.

Nearly 370 years later, *Voyager 1* gave us a close-up look at the Galilean moons. The moons are shown so that you can compare their sizes. Io is slightly larger then our own moon. Ganymede is larger than the planet Mercury. Europa is the dark-surfaced moon. Callisto is the light-surfaced moon.

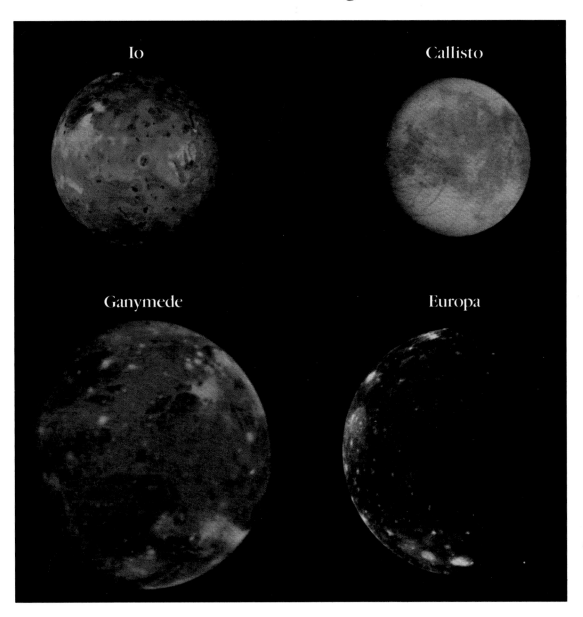

Io [EYE-oh] has something that no other moon in the Solar System has: active volcanoes. The volcanoes erupt often, sending out flows of hot, liquid sulfur. The sulfur changes color as it cools. The surface of Io is always changing with each new volcanic eruption. The black spot in the photograph below is the crater of a dead volcano.

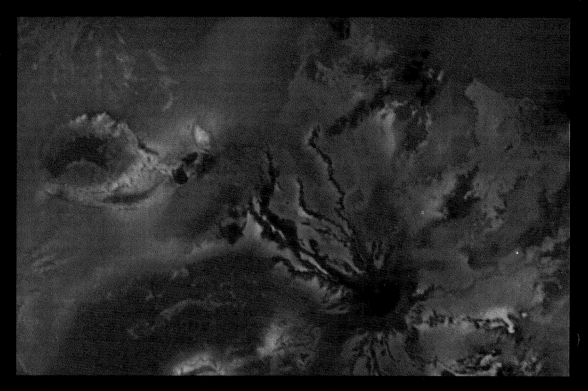

The beautiful photo at right shows an erupting volcano on Io. The blue colors above the volcano show the gas and dust being flung hundreds of miles into space. The volcano was named Loki, after the mischievous god in Norse myths.

Europa [you-RO-pa] is slightly smaller than our moon. From a distance it looks like a smooth, whiteish ball with no high mountains and few of the large craters there are on our moon. In fact, Europa may be the smoothest object in all of the Solar System.

Europa's surface is made of frozen water. This ocean of ice is probably forty to sixty miles thick. Close-up pictures show a network of dark streaks, three to forty miles across. Some of the longer streaks stretch for hundreds or thousands of miles. Many dark patches also can be seen on the surface. In some places bright lines cross the darker streaks and patches.

What caused all these markings? Scientists think that the streaks are shallow valleys filled in with dark material that has been pushed up from inside Europa. The patches may be dark rocks that mixed with water and then froze. And the bright lines are probably low ridges of ice.

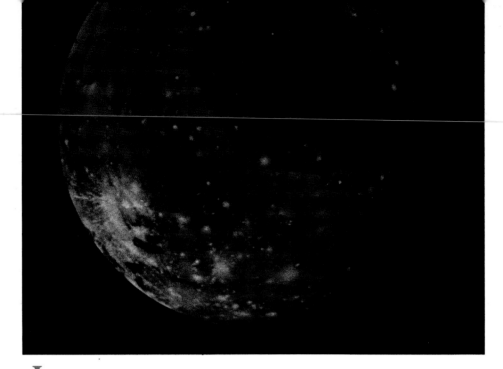

Jupiter's biggest moon is Ganymede [GAN-a-meed]. It is about two-and-one-half times the size of our moon. The white places you see in the photographs are mostly ice. The dark places are mostly rocky soil.

At Ganymede's great distance from the sun, ice won't melt even in full sunlight. But millions of years ago Ganymede was very hot beneath its surface. The heat caused volcanoes to erupt. The volcanic heat melted the ice and it turned to water. The water then seeped under the surface and froze again.

The surface of Ganymede is very wrinkled. The thin white lines are long, narrow hills or ridges of ice mixed with dust and rock. Dark grooves or valleys lie between the ridges. The icy ridges are a few hundred feet high, several miles wide, and hundreds of miles long.

What caused the ridges? Some scientists think that long ago when Ganymede was very hot inside, the heat expanded the surface and cracked it. Water from inside flowed out and filled up the cracks. Then the water quickly froze into icy ridges.

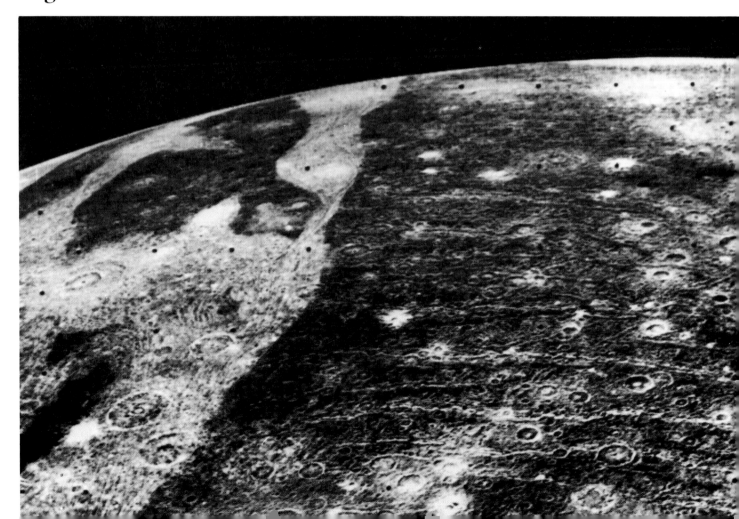

Callisto [ka-LISS-toe] is the outermost of the Galilean moons. It is mostly ice with only small amounts of rock on top of a deep, frozen ocean. At its center is a rocky core.

More than any other of Jupiter's moons, Callisto is covered with large craters. Craters are holes in the ground with ring-shaped walls around them. The craters were formed when meteors, giant rocks from space, crashed into Callisto, melting the surface ice. Like rings of water made when you drop a rock in a pond, slushy waves spread for hundreds of miles and then froze fast enough for the waves to remain.

The outer ring of the crater in the photograph is about fifteen hundred miles across.

When the planets formed, more than four billion years ago, meteors bombarded the surfaces of all of Jupiter's moons. Callisto is the only one where most of the craters have remained. On Io, many of the early craters have disappeared because volcanoes covered the surfaces with melted rocks. On Europa and Ganymede, ice covered most of the craters. But on Callisto, little has happened since the craters were formed.

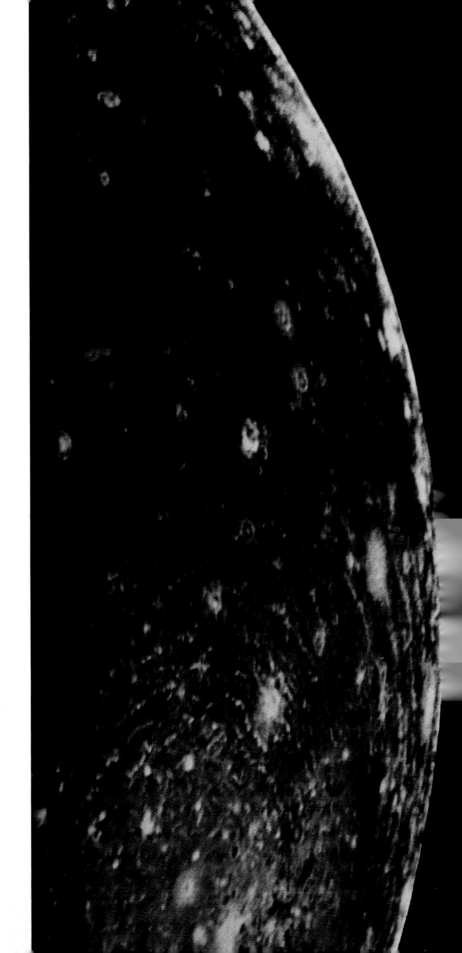

One of the most exciting discoveries of *Voyager 1* was that Jupiter has a thin ring circling the planet. It may have formed from the dust thrown out by Io's volcanoes. No one knows for sure. Jupiter is the third planet in our Solar System shown to have rings, along with Saturn and Uranus.

This photograph shows Jupiter and part of the ring (right). It was taken in Jupiter's shadow on the far side of the planet. Sunlight, coming through the ring from behind, is scattered by the tiny bits of dust in the ring. This is like a puff of dust in a darkened room; you can only see the tiny particles if a ray of sunlight comes through the window.

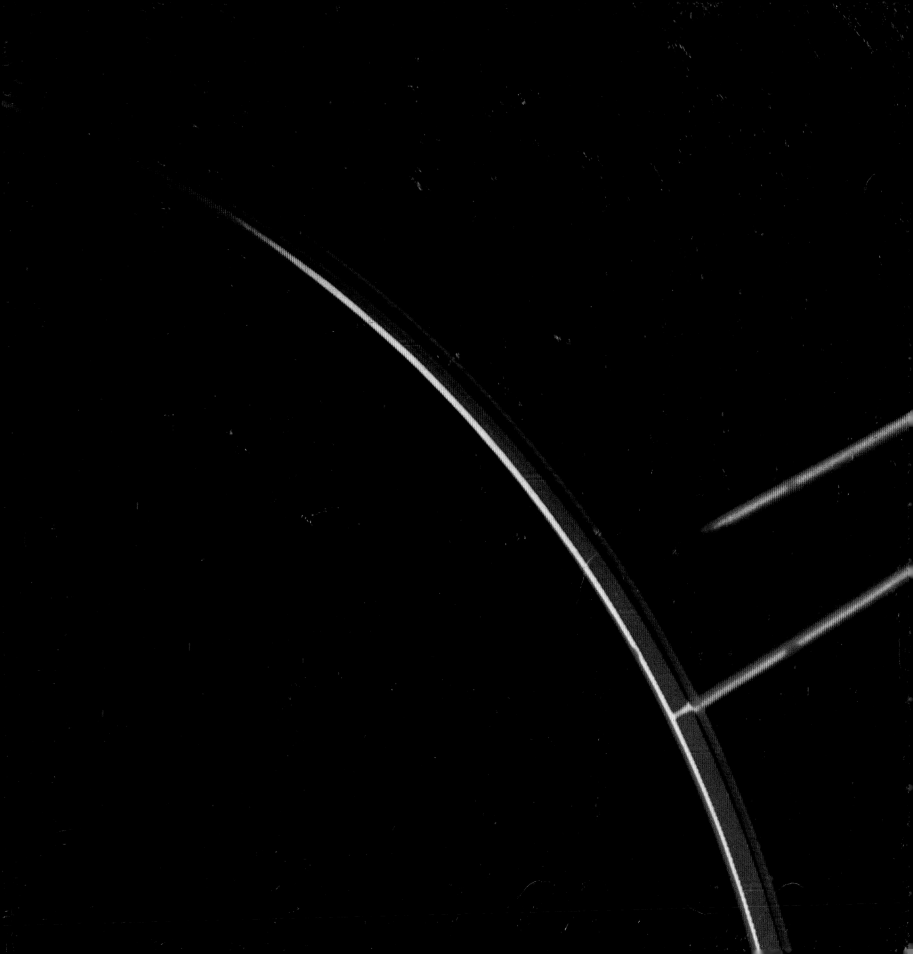

Jupiter is a strange and mysterious world. It is hard to imagine a place more unlike our own planet Earth.

Is there life on Jupiter? No one knows. But life as we know it could not exist. In fact, no spaceship we have could survive its enormous storms. Jupiter remains an alien wilderness in the Solar System, one we have only begun to explore.